Social Software and National Security:
An Initial Net Assessment

Mark Drapeau and Linton Wells II

Center for Technology and National Security Policy

National Defense University

April 2009

Dr. Mark Drapeau is an Associate Research Fellow at the Center for Technology and National Security Policy at the National Defense University. Prior to this, he was an American Association for the Advancement of Science (AAAS) Science and Technology Policy Fellow in National Defense and Global Security at the Department of Defense, and a National Institutes of Health (NIH) Ruth L. Kirschstein Postdoctoral Research Fellow in neurogenomics at New York University. He was also a member of the International Honeybee Genome Sequencing Consortium, within which he analyzed a family of genes underlying complex insect social behavior. Dr. Drapeau holds a B.S. in biology from the University of Rochester, and a Ph.D. in ecology and evolutionary biology from the University of California, Irvine. He can be reached via email at drapeaum@ndu.edu.

Dr. Linton Wells II is a Distinguished Research Professor at the Center for Technology and National Security Policy, and serves as the Force Transformation Chair at the National Defense University. Prior to coming to NDU he served as the Principal Deputy Assistant Secretary of Defense (Networks and Information Integration), the Acting DOD Chief Information Officer, and the Principal Deputy Assistant Secretary of Defense (Command, Control, Communications and Intelligence). Dr. Wells holds a B.S. in physics and oceanography from the U.S. Naval Academy, and an M.S. in mathematical sciences and a Ph.D. in international relations from The Johns Hopkins University. He is also a graduate of the Japanese National Institute for Defense Studies in Tokyo.

Defense & Technology Papers are published by the National Defense University Center for Technology and National Security Policy, Fort Lesley J. McNair, Washington, DC. CTNSP publications are available at http://www.ndu.edu/ctnsp/publications.html.

Contents

Executive Summary

Social software connects people and information via online, informal Internet networks. It is appearing in increasingly diverse forms as part of a broad movement commonly called *Web 2.0*. Resulting social connections are typically serendipitous and can bring unexpected benefits. New social software technologies offer organizations increased agility, adaptiveness, interoperability, efficiency and effectiveness. Social software can be used by governments for content creation, external collaboration, community building, and other applications.

The proliferation of social software has ramifications for U.S. national security, spanning future operating challenges of a traditional, irregular, catastrophic, or disruptive nature. Failure to adopt these tools may reduce an organization's relative capabilities over time. Globally, social software is being used effectively by businesses, individuals, activists, criminals, and terrorists. Governments that harness its potential power can interact better with citizens and anticipate emerging issues.

Security, accountability, privacy, and other concerns often drive national security institutions to limit the use of open tools such as social software, whether on the open web or behind government information system firewalls. Information security concerns are very serious and must be addressed, but to the extent that our adversaries make effective use of such innovations, our restrictions may diminish our national security.

We have approached this research paper as an initial net assessment of how social software interacts with government and security in the broadest sense.[1] The analysis looks at both sides of what once might have been called a "blue-red" balance to investigate how social software is being used (or could be used) by not only the United States and its allies, but also by adversaries and other counterparties. We have considered how incorporation of social software into U.S. Government (USG) missions is likely to be affected by different agencies, layers of bureaucracy within agencies, and various laws, policies, rules, and regulations. Finally, we take a preliminary look at questions like: How should the Department of Defense (DOD) use social software in all aspects of day-to-day operations? How will the evolution of using social software by nations and other entities within the global political, social, cultural, and ideological ecosystem influence the use of it by DOD? How might DOD be affected if it does not adopt social software into operations?

In the process, we describe four broad government functions of social software that contribute to the national security missions of defense, diplomacy, and development. The

---

[1] See: "Net Assessment: A Practical Guide," by Paul Bracken, 2006, *Parameters,* http://www.carlisle.army.mil/usawc/parameters/06spring/bracken.pdf.

first function is **Inward Sharing**, or sharing information within agencies. This includes information sharing not only during military operations, but also within offices for budgets, human resources, contracting, social, and other purposes, and coordination between offices and other units of an agency.

The second function is **Outward Sharing**, or sharing internal agency information with entities beyond agency boundaries. Outward sharing includes coordination during the Federal interagency process; sharing information with government, law enforcement, medical emergency, and other relevant entities at state, local, and tribal levels; and collaboration with partners such as corporations, non-governmental organizations (NGOs), or super-empowered individuals (billionaires, international CEOs, etc.).

The third function is **Inbound Sharing,** which allows government to obtain input from citizens and other persons outside the government more easily. Inbound Sharing includes gauging public sentiment on issues in real time (not unlike instant polling), allows government to receive input on current topics of interest, empowers the public to vote or otherwise give weight to other people's opinions to reach some consensus or equilibrium about online discussions about government issues, and provides a mechanism for crowdsourcing, which is effectively outsourcing projects to a group of people whose membership is not predefined (not unlike a contest or challenge).

The fourth function is **Outbound Sharing,** whose purpose is to communicate with and/or empower people outside the government. This includes a range of efforts such as focused use of information and communications technology (ICT) during stabilization and reconstruction missions, connecting persons in emergency or post-disaster situations, and communicating messages in foreign countries as part of public diplomacy efforts. It also includes functions like using multimedia and social media for better communication with citizens as part of public affairs.

Social software, if deployed, trained on, monitored, managed, and utilized properly, is expected to yield numerous advantages: improve understanding of how others use the software, unlock self-organizing capabilities within the government, promote networking and collaboration with groups outside the government, speed decisionmaking, and increase agility and adaptability.

Along with the accrual of positive benefits, incorporating social software into day-to-day work practices should also decrease the probability of being shocked, surprised, or out-maneuvered. Whether it is misinformation about U.S. actions overseas being spread through new media channels, or new forms of terrorist self-organization on emerging social networks, experimenting with and understanding social software will increase USG abilities to deal with complex, new challenges.

Because social software can add significant value to many ongoing missions, and because citizens, allies, and opponents will use it regardless, this paper recommends that national security institutions, particularly DOD, embrace its responsible usage. While the focus of this paper is on USG national security institutions, many of the conclusions apply to government generally—what many people call "e-Government" or "Government 2.0" — and although there is more to Government 2.0 than social software usage by government entities, this research paper represents a significant advancement towards a strategic understanding of the topic matter.

1. The Web 2.0 Information Technology Revolution

What we now call the Internet and the Web stemmed from national security interest in a resilient, professional computer data-sharing and storage network envisioned by DOD. Not surprisingly, the first web pages were static and unidirectional spaces where owners posted content for readers to observe. In many cases, the owners were large companies and governments generally wishing to push information out to the world.

Now, the Web is graphical, hyperlinked, dynamic, and empowering. Web 2.0 is dynamic and participatory, where software interacts among many users and across many devices, and persons effortlessly shift between author and audience states.[2] Site owners typically offer constantly changing content, increasingly joined in "mashed up" groupings drawn from multiple sources. These may include long original essays, mainstream news headlines, geospatial information, external niche blogs,[3] microblogging[4] conversations, and advertisements tailored to viewers' interests.

More profoundly, readers—even detractors—can alter owner's pages, often by leaving comments, rating the value of the pages, linking related items, or using other mechanisms. New technologies are developed and used in this space very quickly, and the rate is accelerating faster than many realize.

Importantly, in the Web 2.0 world, interactions commonly are multi-directional, interactive, and iterative. An online newspaper reader can comment on an op-ed, and the author can respond—what previously seemed like insurmountable barriers between writers and other public persons has to a large extent melted away. (As many people have commented, "the gatekeepers are dead.") Sometimes even supplementary connections and discussions from these engagements lead to more news than the original writing,[5] and may generate opportunities for participants to educate the original authors.[6]

---

[2] Web 2.0 is a loosely used term with various meanings. For the purposes of this paper, Web 2.0 as defined by the person who coined the term, Tim O'Reilly: "Web 2.0 is the network as platform, spanning all connected devices; Web 2.0 applications are those that make the most of the intrinsic advantages of that platform: delivering software as a continually-updated service that gets better the more people use it, consuming and remixing data from multiple sources, including individual users, while providing their own data and services in a form that allows remixing by others, creating network effects through an architecture of participation, and going beyond the page metaphor of Web 1.0 to deliver rich user experiences."

[3] A blog is the common term for a weblog, a web site with regular entries of commentary, descriptions of events, or other multimedia, often in reverse chronological order. As of late 2008 there were well over 150 million blogs, and growing, although precise numbers are difficult to obtain.

[4] Microblogging, more accurately but less commonly called microsharing, refers to very brief (typically, 140 characters or less) amounts of information shared in a broadcast instant message (IM) format. Twitter.com is the exemplar microblogging social software platform.

[5] The 2006 Virginia Senate race between incumbent George Allen and challenger Jim Webb includes an example of a story that was kept alive in the blogosphere until it was picked up by the mainstream media. The infamous YouTube video of Allen calling a Webb staffer a "macaca," http://www.youtube.com/watch?v=r90z0PMnKwI, viewed by hundreds of thousands of people, led to a

Social software can be thought of as the symbolic antithesis of traditional stand-alone information technology (IT) systems, because it is explicitly designed to share information with other software in other locations (see figure 1 for examples of this transformation).[7] Social software is alternately called Web 2.0 technology, social media, the social web, social technology, and so on, but the term *social software* is the broadest construct that encompasses the topics of interest here. Social software is used here in the most expansive sense—applications that inherently connect people and information in spontaneous, interactive ways.

Social software includes a diverse set of tools, summarized non-exhaustively in table 1. These tools are often grouped into broad categories: personal social networks (Facebook), blogs (WordPress), microblog (Twitter), audio (BlogTalkRadio), video (YouTube), collaborative tools (GoogleDocs), wikis[8] (TWiki), and so on. The fact that many of these communications tools are cheap to make and have relatively short development times and shelf lives makes it difficult for government to keep up with information about new products and trends.

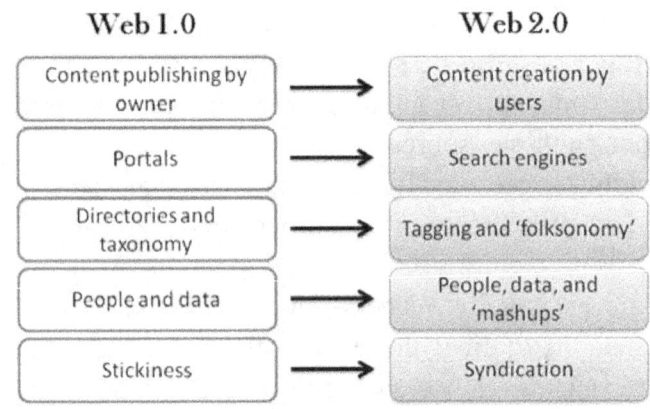

Figure 1: Some differences between features of Web 1.0 and Web 2.0 (non-exhaustive)

Social software is an increasingly important part of global information flow. Wikipedia has about four million articles.[9] YouTube has over 100 million videos.[10] There are more

---

five-minute interrogation of Allen on *Meet the Press,* http://www.youtube.com/watch?v=wRfP3vj8Gl8. Allen lost to Webb.

[6] Dan Gillmor, *We the Media,* http://www.authorama.com/we-the-media-1.html.

[7] Figure adapted partially from p. 16 of, *Leveraging Web 2.0 in Government,* by Ai-Mei Chang and P.K. Kannan, University of Maryland, for the IBM Center for the Business of Government, http://www.businessofgovernment.org/pdfs/ChangReport2.pdf.

[8] Wikis are pages or collections of pages designed to enable anyone accessing them to contribute or modify content easily. Wikis are often used to create collaborative websites and to power community websites, http://en.wikipedia.org/wiki/Wiki.

[9] Wikipedia, launched in 2001, is a free, multilingual, volunteer user-generated, collaborative encyclopedia with over 12 million articles. Most any article can be edited by most any person. The word "Wikipedia" is part "wiki" and part "encyclopedia." Wikis are only a subset of online collaborative formats. See, for example, this study of collaborative paper writing, Christopher King, 2007, "Multiauthor Papers Redux: A New Peek at New Peaks." http://archive.sciencewatch.com/nov-dec2007/sw_nov-dec2007_page1.htm.

[10] YouTube, now a subsidiary of Google, is a video sharing website with user generated and uploaded video clips, free views, and encouraged sharing via embedding into other websites like blogs. Most content

than 200 million blogs on the Internet. Second Life, a virtual reality world,[11] has over 1.5 million "residents" and generates more than a million dollars a day in real world transactions.

Although social software tools are sometimes dismissed as "time wasters" or "programs for kids," an informal survey we conducted of friends and colleagues working in science and technology positions around the government revealed that, more often than not, popular Web 2.0 sites are not blocked on office computers. For example, we found that Facebook was blocked in some parts of the Departments of Defense, Energy, Education, Agriculture, and Health and Human Services, but was available at the Agency for International Development, the National Science Foundation, the National Oceanographic and Atmospheric Administration, the Environmental Protection Agency, the State Department, and both houses of Congress. Other Web 2.0 sites like the microsharing platform Twitter and the video sharing site YouTube were even less restricted.

Communication on the Internet is no longer a controlled, organized, exclusive, product-driven monologue; it is an authentic, transparent, inclusive, user-driven dialogue. Increasingly, people who habitually use the Internet are not only browsers or readers, but also providers and participants. And listening is the new talking. If you work for the government, someone—right now—is talking on the Internet about your agency and your mission—effectively, your brand.[12] The people participating in these conversations have less trust in mainstream media messaging and traditional advertising, and more trust in word-of-mouth conversations within their social networks. Government ignores this fact at its peril.

Use of social software as ICT is creative and collaborative. Large corporations conduct market research by monitoring public sentiment about their products. Small businesses use blogs and other interactive media to identify new customers or advertising opportunities. Organizations of all kinds recruit in targeted niches. Empowered individuals build personal brands for their hobbies, and turn those hobbies into professions. Authors put unfinished works online to obtain early feedback from their biggest fans. Members of the media solicit interview questions from their audience in real time. In all cases, the activities involve important behavioral constructs: enabling, inspiring, listening, engaging, and influencing.

---

is uploaded by individuals, although increasingly companies, especially media-related ones, are also offering content. There are, additionally, examples of official USG YouTube use.

[11] Virtual reality is technology that allows users to interact with computer-simulated environments via one or more senses (usually visual with possible others). These realities have different degrees of realism. They are often used in online games and increasingly in military and other kinds of training, http://en.wikipedia.org/wiki/Virtual_reality.

[12] Mark D. Drapeau, "Government 2.0: What's Your Brand?," http://mashable.com/2008/09/03/government-brand/.

Category	Example 1	Example 2	Example 3
blogging	Blogger	Wordpress	TypePad
microblogging	Twitter	Pownce	Plurk
wikis	Pbwiki	WetPaint	WikiDot
social networks	Facebook	LinkedIn	Plaxo
bookmarking	Delicious	ma.gnolia.com	Fark
aggregators	Digg	StumbleUpon	FriendFeed
photos	Flickr	Picasa	Photobucket
audio/video	YouTube	Blip.tv	Hulu
messaging	Gchat	AIM	Yahoo! Messenger
Twitter applications	Twhirl	Tweetdeck	TwitterBerry

Table 1. Common Social Software Tools, by Major Function/Category

Private-sector social software thought leaders are optimistic that the USG will make increased use of social software.[13] Tim O'Reilly, the social technology publisher and evangelist who coined the term *Web 2.0*, stated in his Obama for President endorsement that, "there are efforts already underway to build better tools for two-way communication, for government transparency, and for harnessing innovations from outside the public sector to improve the work of the public sector."[14] Indications from the Obama for President Campaign, the Presidential transition, and the early days of the Obama Administration are that this will continue.[15]

While DOD has shaped many developments in ICT over the decades via internal and external research, social software has largely been developed without public funding by companies of modest size for commercial applications. The microsharing platform

---

[13] See, for example, the event held by the New America Foundation and Google called, Wiki White House, http://www.newamerica.net/events/2008/wiki_white_house, and Dan Froomkin for the Huffington Post, "It's Time for a Wiki White House," http://www.huffingtonpost.com/dan-froomkin/its-time-for-a-wiki-white_b_146284.html. Some interesting examples pertaining to national security can be found in *Mashup the OODA Loop*, June 2008, MITRE Technical Report MTR070365.

[14] Tim O'Reilly, "Why I Support Barack Obama," http://www.huffingtonpost.com/tim-oreilly/why-i-support-barack-obam_b_139058.html.

[15] Edelman, the public relations firm, published an excellent review, "The Social Pulpit: Barack Obama's Social Media Toolkit," http://www.edelman.com/image/insights/content/Social%20Pulpit%20-%20Barack%20Obamas%20Social%20Media%20Toolkit%201.09.pdf, that deals with social tools used during his campaign. The "Organizing for America" initiative (announced by President Obama on YouTube) seems poised to continue hearing citizen voices on important issues, http://www.techpresident.com/blog/entry/33581/organizing_for_america_launches_structure_tbd. President Obama also made news by insisting on keeping his PDA while in the White House, http://seattletimes.nwsource.com/html/nationworld/2008661443_blackberry23.html. Finally, see the White House memo on Transparency and Open Government, http://www.whitehouse.gov/the_press_office/TransparencyandOpenGovernment/.

Twitter, for example, was invented by a handful of young professionals in Silicon Valley as a way to broadcast to people where good parties or other social events were in real time, like a text-based CB radio. Today its users include senior government officials, major mainstream media figures, and famous athletes and movie stars, and applications range from measuring brand sentiment to emergency relief to raising money for charities around the world.

There is no coordinated, department-wide policy for DOD (or, insofar as we are aware, any other USG agency) or set of guidelines for using the universe of social software tools internally, between agencies or other entities, or with the public. It is unclear in many cases who, what, when, where, why, and how such tools should be used while at work, and while not at work. This leads to confusion and inconsistencies. One USG agency blocks a certain Web 2.0 site, and another down the street allows it.

There are pockets of progressiveness throughout the USG, and empowered individuals are experimenting with these tools, and finding workarounds when sites are blocked (for example, accessing YouTube from a personal laptop with a wireless connection, or from an iPhone), often without official guidance.[16] Indeed, there have been numerous social software experiments throughout the USG, some of them relatively high profile.[17] People with knowledge of how government works, technical understanding of social software, and an interest in public service have been informally dubbed the "Goverati."[18]

Despite some limited success, isolated pockets of bottom-driven informal pilot projects are not the same as a coordinated top-down effort to determine appropriate government uses for social software. Such broad uses include balancing security with transparency, writing policies for use of social software, training personnel to be ready to use the tools, conducting research and acquiring private sector materiel as appropriate, understanding its uses for intelligence and public affairs applications, and assessing the strategic implications for the USG and other countries.

---

[16] Mark D. Drapeau, Government 2, "Being Individually Empowerful," http://mashable.com/2008/08/26/government-20-being-individually-empowerful/.
[17] The independent Government 2.0 Best Practices Wiki, http://government20bestpractices.pbwiki.com/, is one informal, non-exhaustive directory of such efforts. For a collection of best practices in the private-sector pertaining to large companies using blogs, see http://blogcouncil.org.
[18] Mark D. Drapeau for *Federal Computer Week*, "The rise of the Goverati," http://www.fcw.com/Articles/2009/02/23/drapeau-rise-of-goverati.aspx.

2. Framework: Government Social Software Functions

Here we define four broad functions of social software relevant to national security—Inward Sharing, Outward Sharing, Inbound Sharing, and Outbound Sharing. The functions can be diagramed along two axes (see figure 2). The *x* axis deals with whether social interaction is done largely between known individuals (e.g., Bob sends information to Jill, Jackie sends information to a small office of four persons, and so forth), or if individuals are interacting with groups of relatively unknown persons (e.g., a research office solicits ideas from any U.S. citizen with an advanced engineering degree, a public affairs office records a video blog

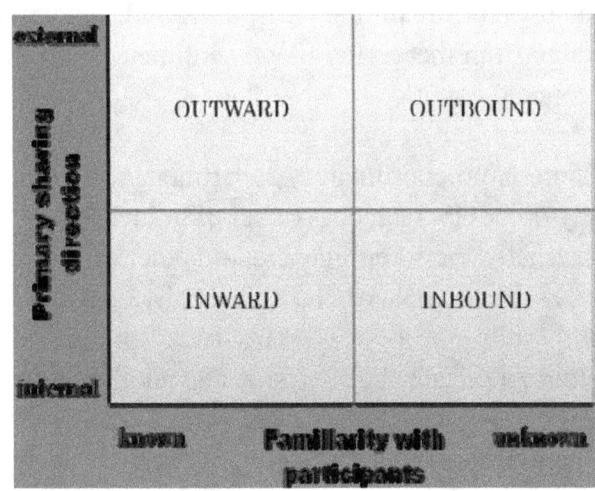

Figure 2: Four-quadrant government social software framework

about a new policy and invites open feedback on its website). The *y* axis represents the direction of sharing—whether the user information is mainly being sent out, or accepted in. We note that these quadrants do not represent truly discrete categories, but rather a continuum of states along which social software applications lie.[19]

For each of these four functions of social software in government, we give current and successful examples from both the public and private sectors. While there is some overlap between the four functions, and some examples can reasonably apply to more than one function, the structure is nevertheless useful for conceptualizing a larger framework of how social software fits into DOD and USG missions and goals. The focus of this paper is on national security applications, and specifically on DOD, but much of it can be generalized and may serve as a framework for other parts of the USG, and even state and local governments.

2.1 Function One—Inward Sharing

The first general function of social software in government (DOD will be used throughout this paper, though many platforms, tactics, anecdotes, and strategies apply to

---

[19] For an alternative, and somewhat overlapping framework for Web 2.0 in government, see figure 4 in IBM's report http://www.businessofgovernment.org/pdfs/ChangReport2.pdf - focused more explicitly on relationships between government, businesses, and citizens.

most entities across the USG) is Inward Sharing, or intra-institutional sharing—sharing information within a department.[20] This includes all forms of information sharing, including: as part of military operations, within a lessons-learned process, military intelligence gathering and analysis, human resources decisionmaking, networking warfighters' families, communicating general office information, and coordination of information between two or more offices or enterprises within DOD.[21]

Using social software as a platform for Inward Sharing, employees, contractors, and other trusted stakeholders could form "heterarchies" of decentralized, empowered individuals working on problems within a complex, hierarchical organization like DOD. Trustworthy information, whether stored centrally or distributed, must be discoverable, accessible, and understandable by anyone within the social network. Open sharing provides more "eyes on target" and makes the discovery of weak signals in batches of noise more likely.

One gap in DOD communications architecture that includes radio, video chat, instant messenger services, email, and web portals, is the ability for information to be pushed that neither party "knows" is needed *a priori*. While delivering the right information to the right person at the right time sounds like a good idea, it is remarkably difficult to predict the relevance of information.[22] Many users producing a persistent flow of *possibly* relevant information provide the framework for users to decide the importance of information, an "ambient awareness" of others' behavior,[23] and surprise discoveries of small pieces of knowledge. This kind of Inward Sharing, exemplified in the private sector by the popular service Twitter, has the potential to close information gaps between warfighters during a deployment, between warfighters and civilian planners and analysts of various kinds, and between civilians away from the battlefield in other DOD roles.

Importantly, such Inward Sharing tools not only have conventional uses, like intelligence analysis, but also relatively mundane but very important advantages not considered as frequently, such as social networking for warfighters new to a base to help their families find local schools and commerce, or advertising job vacancies and spreading morale-boosting messages within offices. Encouraging this kind of knowledge interaction in which people are more easily connected with each other will encourage open discussion, community building, and efficiencies of scale. Finally, social software may even promote frugality—phone calls, focus groups, and airline tickets can be expensive, and new technology can help circumvent these costs in some situations.

---

[20] See, for example, the Forrester Research report, http://www.forrester.com/Research/Document/Excerpt/0,7211,43882,00.html.

[21] Applied Minds, Inc. is currently building a Facebook-like social networking platform for the Air Force Research Laboratory, http://peoplepointsystems.com/Aristotle_NAECON_PDF.pdf.

[22] Taleb, N. N. (2007), *The Black Swan: The Impact of the Highly Probable*. New York, NY: Random House.

[23] *Ambient awareness* is a term popularized by Clive Thompson in "Brave New World of Digital Intimacy" in the *New York Times*, http://www.nytimes.com/2008/09/07/magazine/07awareness-t.html. It describes knowing people's behavioral habits coming from reading lots of tiny bits of information flow about them.

Example 2.1.1 (inside DOD, at Army): CompanyCommand. CompanyCommand started as a small, grass roots effort to connect U.S. Army captains, who command at the company level (~150 soldiers) directly. The notion was to pass knowledge directly and more efficiently without going up through the chain of command to a more senior officer and then have it disseminated back down through the brigade (or other relevant unit) hierarchy. This was outside policy at the time. When discovered by senior decision makers the grassroots, voluntary forum was threatened with being shut down. It was eventually deemed so useful that it is now hosted on military, and not private, computer servers at http://companycommand.army.mil/. Similar functionality now exists for other levels of command.

Example 2.1.2 (outside DOD, at IC): INTELINK. INTELINK, run by the Office of the Director of National Intelligence (ODNI) is a sophisticated internal government social network that includes Intellipedia and other useful social software tools that are similar to commercial sites like Wikipedia and YouTube.[24] This can be a tremendously powerful system for sharing intelligence, particularly raw intelligence reports, thoughts, photos, and so forth, insofar as people use it. Having different but connected systems at various levels of security classification is also very useful. ODNI has also developed a government-wide (in principle) enterprise email system with the domain ugov.gov, for "unclassified government" email. The password-protected ugov.gov acts not only as a central phonebook of sorts, but also as a "passport" to access and edit parts of INTELINK.

Example 2.1.3 (outside DOD, at DHS): TSA Idea Factory. In 2007, the Transportation Security Administration (TSA), part of the Department of Homeland Security (DHS) launched IdeaFactory, an internal collaboration tool designed to tap into collective wisdom, acting as an electronic suggestion box of sorts. Ideas to improve TSA are submitted, and employees vote them up or down; highly ranked ideas are considered by TSA leadership. Within about a year, approximately 4,500 ideas were submitted and about 20 implemented. The inexpensively built community is lightly edited and largely self-policing.

2.2 Function Two—Outward Sharing

The second function is Outward Sharing, or inter-institutional sharing—sharing internal agency information with entities outside departmental boundaries. Outward Sharing includes various kinds of coordination and collaboration during the formal and informal Federal interagency process. It also encompasses sharing USG information with

---

[24] The idea for such a system stems from "The Wiki and the Blog: Toward a Complex Adaptive Intelligence Community," D. Calvin Andrus, Central Intelligence Agency, *Studies in Intelligence*, Vol. 49, No 3, September 2005, http://papers.ssrn.com/sol3/papers.cfm?abstract_id=755904.

government, law enforcement, medical emergency, and other relevant entities at state, local, and tribal levels. Finally, Outward Sharing facilitates collaboration with USG partners such as large corporations, non-governmental organizations (NGOs), or super-empowered individuals.

The 2005 natural disaster of Hurricane Katrina is a textbook example of the need for multi-agency, multi-government, multi-media engagement in an ad hoc and constantly evolving manner. More recently, people using social software have been able to make useful contributions during flooding in Bangladesh, the California wildfires, and Hurricane Gustav.

Example 2.2.1 (public-private): GovLoop. The online social network called GovLoop (http://www.govloop.com/) was created to convene and informally share information among government employees and contractors through information sharing on personal profiles, blogs, etc. GovLoop was designed by a DHS ICT professional in his spare time using a simple platform called Ning. GovLoop effectively circumvents the fact that the USG does not itself have an internal social network like people are familiar with in their personal lives.[25] Now with over 7,000 users from Federal, state, and local governments, GovLoop has become an increasingly popular way for people to blog, network, and learn on a professional and also personal level. The online social network has facilitated in-person get-togethers such as the recent 500-person Government 2.0 Camp in Washington, DC, and has many other applications, including organizing formal USG events, disseminating critical information, and conducting informal employee polls about benefits or social activities. Postings and subsequent discussions on a system like GovLoop may replace many FYI emails that are sent every day.

Example 2.2.2 (public-private): S TAR-TIDES. The international knowledge-sharing research project known as STAR-TIDES[26] promotes unity of effort among diverse organizations where there is no unity of control, as there often will not be among the disparate and rotating entities engaged in complex operations.[27] It leverages a global

---

[25] Anthony Williams, co-author of *Wikinomics*, commented on a Sept 19, 2008 panel at the National Defense University that, "The Ontario government blocked Facebook, so everyone moved to MySpace. It's a futile exercise." This is taken to mean that if people at a company, the government, etc. want some functionality (social network, chat, microsharing, etc.) they will find a way to circumvent the rules to get it. Security officers and senior leaders need to understand this and look for ways to enable responsible use, rather than imposing blanket prohibitions on mission-essential functions. There is ample evidence that the inevitable workarounds will create their own security vulnerabilities, leaving everyone worse off than a collaboratively derived solution would have.

[26] TIDES stands for Transportable Infrastructures for Development and Emergency Support. It is part of a broader research effort called STAR - Sustainable Technologies, Accelerated Research, http://star-tides.net'.

[27] The definition of complex operations has changed over time—sometimes including combat, sometimes excluding it, sometimes encompassing disaster relief, sometimes not, and usually focusing only on missions overseas. For example, the Center for Complex Operations website states that "stability operations, counterinsurgency and irregular warfare [are] collectively called 'complex operations." A

social network using, among other things, social wikis, online photos, video, and microblogging tools like Twitter. STAR-TIDES seeks to enhance the ability of civilian coalitions to work in stressed environments (post-war, post-disaster, impoverished) and extend the military's ability to work with them. The overall objective is to connect people who have problems with those who may have solutions, and to save money through cost-effective logistic solutions and by dividing supply responsibilities among different providers. As informal covenants (handshakes, distributed data storage, Web 2.0, etc.) increasingly replace traditional kinds of formal agreements among entities that must work with DOD, social software and networks like STAR-TIDES that are enabled by it will become increasingly important.

Example 2.2.3 (public-private): NIUSR5. The National Institute for Urban Search and Rescue, Readiness, Response, Resilience, and Recovery (NIUSR5) (http://www.niusr.org/), based in Santa Barbara, CA has recently decided to network its members and share information through tools available via the free, popular social networking site LinkedIn.[28] These include user groups, search, email digests, discussion threads, and person-to-person messaging. All information is contained within the LinkedIn system, which lets it serve as a makeshift "enterprise" ICT system. To supplement this communication, particularly in mobile, disaster relief situations, NIUSR5 also has decided to make use of Twitter. People can more easily be connected with authorities, form ad hoc social networks, and so forth.[29]

2.3 Function Three—Inbound Sharing

Inbound Sharing allows government to obtain input more easily from citizens and even persons outside the country when appropriate. This third function of social software includes gauging public sentiment on issues in real time (not unlike polling), allows government to receive personal input on current topics of interest (perhaps even blunt and anonymous input), empowers the public to vote as part of online discussions relevant to government issues, and provides a mechanism for crowdsourcing, which is effectively

---

forthcoming book, *Civilian Surge*, edited by Hans Binnendijk and Patrick Cronin, adopted a more expansive definition that includes humanitarian assistance and disaster relief, at home and abroad, which is the one used in this paper, http://www.ndu.edu/ctnsp/pubs/Civilian%20Surge%20DEC%202008.pdf.
[28] LinkedIn.com is designed as a business-oriented social networking site with over 30 million users representing well over 100 industries and areas. Various tools like resume posting, friend making, question posing and discussion, and internal message sending allow Web 2.0-style professional networking using social software.
[29] In the way of an example, InSTEDD (Innovative Support to Emergencies, Diseases, and Disasters), http://instedd.org, whose leadership has been closely associated with NIUSR, is developing social software to solve human interaction problems and "collaboration gaps" in emergency situations where access to the Internet is slow or unreliable.

outsourcing projects to a group of people that is not predefined (not unlike a contest or challenge).[30]

The government can play a large role in providing people who are already having conversations about topics with a platform where they can network and share information with Washington, DC, (or at other levels) about issues important to them—education, health, military action, etc. This information is extremely valuable. Moreover, social software can more easily facilitate challenges. The Defense Advanced Research Projects Agency is well-known for its DARPA Grand Challenge, which challenges entrants to design autonomous vehicles meeting a strict set of standards.[31] Smaller problems can be solved in a similar manner, using social tools to identify people who might have a proclivity for the topic, advertising new challenges, celebrating winners, and so on.

DOD collects, maintains, and publishes a tremendous amount of data, yet much of it is out of print, difficult to find, or hard to work with because of security classification issues, file formats, lack of data tagging, and so forth. Yet much of it is also is unique and could be useful to other government employees, professors and researchers, and the average citizen. Additionally, private sector firms and empowered individuals may very well devise interesting manipulations and mashups of DOD information that would be of use to DOD itself.

Clearly, such initiatives involving the public must be accompanied by careful policy and legal oversight, but some experts think that organic social networks could replace or complement government "market research" in many situations. This form of outreach could also save money and increase efficiency. In essence, social software can be used to build online work environments that focus on and enhance collaboration, and to some degree redraw divisions of labor in society.[32]

Example 2.3.1 (outside DOD, at Obam a Transition): Change.gov site. The website Change.gov was the official site The Office of the President-Elect from November 2008 to January 2009. One section of the site, "Open for Questions," allowed people to submit questions to the President-Elect and/or vote on the relative importance of or interest in

---

[30] *Crowdsourcing* is the act of taking a task traditionally performed by an employee or contractor and outsourcing it to an undefined, generally large group of people in the form of an open call for ideas; See: Jeff Howe, *Crowdsourcing: Why the Power of the Crowd is Driving the Future of Business*, http://crowdsourcing.typepad.com/. An excellent non-profit approach to crowdsourcing for counterterrorism is "Force Multiplier for Intelligence: Collaborative Open Source Networks," http://www.csis.org/media/csis/pubs/080204-deborchgraveforcemultiplier.pdf, Report of the Transnational Threats Project, Center for Strategic and International Studies, 2007.

[31] The DARPA Grand Challenge, http://www.darpa.mil/grandchallenge/index.asp, inspires teams whose members have diverse backgrounds and experience to build autonomous vehicles to drive through traffic and meet various benchmarks and qualifications, and in turn solve DOD problems indirectly. The X Prize Foundation challenges work similarly, http://www.xprize.org/.

[32] Comments paraphrased from Anthony Williams (co-author of *Wikinomics*) and Bruce Klein (Vice President of Cisco for Public Sector) during a panel discussion held on Government 2.0 at National Defense University, Sept 19, 2008.

other people's questions.[33] Main topic areas for questions included the economy, national security, foreign policy, education, health care, and energy. According to the website, over 100,000 people submitted over 75,000 questions and cast over 4.7 million votes. Inbound sharing like this can aid decisionmakers when analyzing which projects or programs to emphasize, fund, etc.

Example 2.3.2 (outside DOD, at DC OCTO): Apps for Democracy. In November 2008, the Office of the Chief Technology Officer of the District of Columbia held a contest called "Apps [Applications] for Democracy" where, with few restrictions, anyone could access the District's data on parking meters, crime, potholes, spending, and the like and design applications for any variety of platforms (PC, Mac, mobile phones, etc.) that connected residents with District information more effectively. Entrants, mainly but not exclusively from the DC-VA-MD area, competed for $20,000 in prizes; The Office of the Chief Technology Officer (OCTO) estimated their return on investment (ROI) from the 30-day Apps for Democracy contest was 4000 percent.[34] This was a significant example of a relatively inexpensive venture that tapped into the collective wisdom of the "global brain" and provided great value for constituents.

Example 2.3.3 (private sector): Innocentive. The Massachusetts-based company Innocentive connects companies seeking answers to technical problems (say, Dow Chemical trying to synthesize a compound used in a cleaning product using a novel mechanism) with experts outside the company who nevertheless have expertise and are able to solve the problem. Solvers compete for prizes that vary with the job, and many of the solvers are globally based. Perhaps most significantly, people often do not solve problems in their primary area of expertise—in this model a retired biochemist might work on a neurophysiology problem with a retired electrical engineer friend. Innocentive's business model works because companies have a strong incentive to hire them as a "crowdsourcing firm" for tough technical problems, and users have an incentive to receive financial rewards for putting their knowledge into action in their spare time.

2.4 Function Four—Outbound Sharing

The fourth function of social software in government is Outbound Sharing, whose purpose is to communicate with people outside the government, or empower them to communicate with each other. This function includes a complicated range of USG

---

[33] Just before publication of this paper, the Obama White House launched a similar site, http://www.whitehouse.gov/openforquestions/, for the Administration to take questions for about three days. In March 2009, over 75,000 people submitted over 80,000 questions and cast over 3 million votes.

[34] The force behind Apps for Democracy was DC CTO Vivek Kundra, who is currently the national CIO. Apps for Democracy Yields 4,000% ROI in 30 Days for DC Gov, http://www.istrategylabs.com/apps-for-democracy-yeilds-4000-roi-in-30-days-for-dcgov/. More generally, this is part of a larger movement sometimes termed "Metagovernment." See http://metagovernment.org/wiki/Main_Page.

outreach efforts that include ICT deployment during complex operations such as stabilization and reconstruction missions,[35] connecting persons in emergency or post-disaster situations, and communicating USG messages to foreign countries as part of public diplomacy efforts. It also includes using new Web design techniques, multimedia platforms, and social media to more effectively provide raw government data and information to, and communicate with, citizens as part of a public outreach mission that is complicated by the rapidly evolving face of mainstream media.

The combination of free social software with inexpensive mobile devices or donated computers can empower people to self-organize information-sharing networks that are not bound by Federal, state, local, or many other structural constraints. Social software that operates on simple cell phones or personal digital assistants and incorporates geographical information is becoming ubiquitous. Globally, empowerment through social technology also can be a useful tool for public diplomacy: there are many ways in which government can use social software in a transparent manner to deliver information and have conversations while offering security, trust, and accountability.

Example 2.4.1 (outside DOD, at State): Public Diplomacy. Colleen Graffy, formerly Deputy Assistant Secretary of State for Public Diplomacy, utilized social software, most notably Twitter, to impress her personality and message on foreign media prior to arriving in their countries, and after leaving. Her use of new media prompted a variety of reactions, public discussions, and controversy.[36] Graffy noted that a young Romanian student told her, "We feel like we already know you—you are not some intimidating government official. We feel comfortable talking with you." This students' sense comes from something that social software consultants call ambient awareness—the notion that others get by reading short messages about your life over a period of time; people in Graffy's network—her "followers"—get a sense of whether you wake up early or late, read a lot or a little, work or play hard, and so forth, humanizing you. This humanization can be thought of as softening an audience prior to an initial in-person meeting, and keeping them aware of you after you have left—which is very powerful if done properly.

---

[35] Larry Wentz, Franklin Kramer, and Stuart Starr, "Information and Communication Technologies for Reconstruction and Development: Afghanistan Challenges and Opportunities," NDU-CTNSP Defense and Technology Paper 45, http://www.ndu.edu/ctnsp/Def_Tech/DTP%2045%20Afghan%20ICT.pdf.

[36] Colleen P. Graffy, "A Tweet in Foggy Bottom" (Op-ed), *Washington Post*, December 24, 2008, http://www.washingtonpost.com/wp-dyn/content/story/2008/12/24/ST2008122400049.html; Nathan Hodge, "Diplo-Twittering at the Department of State," *Wired*, Dec 24, 2008, http://blog.wired.com/defense/2008/12/diplo-twitterin.html; Al Kamen, "Live From Iceland, or Maybe Greenland, It's the Dipnote Tweet Show," *Washington Post*, December 10, 2008; Page A23, http://www.washingtonpost.com/wpdyn/content/article/2008/12/09/AR2008120902774.html?nav=rss_opinion/columns; Spencer Ackerman, "Diplomats Use Twitter to Give the World TMI," *The Washington Independent*, December 8, 2008, http://washingtonindependent.com/21346/diplomats-use-twitter-to-give-the-world-tmi; Charles J. Brown, "Dipnote Follies: Twitter TMI, Nukes, and Human Rights," December 10, 2008, http://www.undiplomatic.net/2008/12/10/dipnote-follies-twitter-tmi-nukes-and-human-rights/.

Example 2.4.2 (outside DOD, in Congress): Representative-Constituent Interaction.
Some members of Congress have led the way as advocates for utilizing social software to stay more connected to average citizens, particularly those in their home districts, often far outside Washington, DC. Foremost among these members of Congress is Rep. John Culberson (R-TX), whose experiments have included the cutting edge microsharing platform Twitter and the live Internet video broadcast platform Qik. He has frequently videophone-casted meetings within his office and interviews with the media, and contacted people late into the night on Twitter (one of the authors has personally experienced this). Culberson's technological sophistication prompted rules changes that broaden the scope of media technologies that members can personally use.[37] Moreover, use of social software has arguably increased his profile in Washington, DC—which may have downstream benefits in the workplace.

Example 2.4.3 (private sector): "Comcast Cares."   In the private sector, the telecommunications company Comcast has made very interesting use of Twitter. By monitoring online Twitter conversations in real time via its "Comcast Cares" account, Comcast staff detects complaints (say, about late cable TV installations) and then uses that information to dispatch personnel to the customer or otherwise fix the problem as best as possible, while simultaneously reaching out directly to the customer to inform them of the internal company action via Twitter. This interactive, transparent, and immediate customer service not only works but also adds tremendous positive value to the company via word-of-mouth, the most powerful force in the marketplace. Importantly, what Comcast has implemented is not passive listening, but rather a multi-directional public outreach campaign that more directly connects customer needs with company resources.

---

[37] Andrew Feinberg, Oct 2, 2008, "House Relents on New Media, Adopts Updated Rules for Web Video," http://technosailor.com/2008/10/02/house-relents-on-new-media-adopts-updated-rules-for-web-video/.

3. Challenges to Government Social Software Utilization

Despite these successful examples, there are many obstacles at numerous levels of the hierarchy to moving from isolated pockets of experimentation to a broader policy of using such software. Some cases involve rules established by agencies outside the traditional national security community. For example, the Office of Management and Budget (OMB) has a de facto chief information officer position in its Administrator for e-Government and Information Technology, who has influence on government-wide IT initiatives.

On the whole, the USG is designed to be conservative and hard to change quickly. In addition to the checks and balances built into the American system, this is reinforced by factors such as interagency interactions, classic bureaucracies, security concerns (both legitimate and excessive), infrastructure inflexibilities, employee demographics, etc. Thus, while there may be similarities to the challenges faced by innovative approaches in corporations or other large organizations, the USG has a special set of characteristics and concerns. Most of these issues are not unique to adoption of changes associated with social software, but they are nevertheless valuable to review.[38]

3.1 Interagency Interactions

The many responsibilities of the departments, agencies and other units of the USG range from education improvement to domestic health concerns to overseas military operations. The breadth and importance of the tasks usually requires coordinated efforts from more than one part of the USG, especially in the national security area. But a host of factors encourages stovepiped "cylinders of excellence" and makes interagency collaboration difficult. Recent studies suggest that extraordinary changes in law, organization and processes will be needed to clarify roles and responsibilities and improve performance.[39]

---

[38] While this assessment was being written, an excellent online paper by USG webmasters was published, dealing with day-to-day operational problems in using Web 2.0 tools in government, and offering brief solutions. It is "Social Media and the Federal Government: Perceived and Real Barriers and Potential Solutions," Federal Web Managers Council, http://www.usa.gov/webcontent/documents/SocialMedia Fed%20Govt_BarriersPotentialSolutions.pdf. Another excellent white paper, written by Geoff Livingston of CRT/tanaka for the business community and titled "The Cultural Challenge to Integration," http://www.crt-tanaka.com/documents/The_Cultural_Challenge_to_Integration-Livingston.pdf.

[39] See Project on National Security Reform: Forging a New Shield (2008), http://pnsr.org/data/files/pnsr%20forging%20a%20new%20shield.pdf, and Dr. Anthony Cordesman and Hans Ulrich Kaeser, 2008, Defense Procurement by Paralysis: Costly Mortgages for the Next Administration, CSIS, http://www.csis.org/media/csis/pubs/081114_defense_procurment_by_paralysis.pdf. However, see also the March 19, 2009 memo from National Security Advisor James Jones on "The 21st Century Interagency Process," which outlines a proposal for better interagency interaction with regard to the National Security Council and its members and issues.

3.2 Classic Bureaucracy

Even if the interagency relationships were improved, collaboration is never easy in a bureaucracy, for reasons that have been well documented.[40] Interagency collaboration tools that bring new technology and change social norms may compound as many problems as they fix. Informal survey results show striking inconsistencies across agencies regarding which social software sites can and can not be accessed from work computers—even though many Web 2.0 sites like TripAdvisor.com (travel information and advice), LinkedIn.com (business social network), and Flickr.com (online photos) have legitimate work applications. Inconsistent policies involving the need for social software, the lack of access to it, and inconsistent rules and regulations regarding its personal and professional use lead to confusion and frustration within the USG.

3.3 Information (Mission) Assurance

Though it opens exceptional opportunities, the use of social software also involves security risks. The balance that the USG maintains between information sharing and information security has metaphorically been compared to inhaling and exhaling. A key theme of the 2008 DEFCON hacker convention[41] was how to exploit social software and social networks, primarily as a way to exploit reduced privacy on the Internet, gather information for identity theft, and prepare "custom-tailored, laser-focused attacks."[42] Additionally, social software could be used to promote hoaxes or other false alarms, diverting valuable resources away from real missions in pursuit of ghosts. Finally, there is the issue of authoritative versus unvetted government information—if anyone can say anything, perhaps one warfighter will follow incorrect advice in another warfighter's blog, rather than strictly follow a manual, leading to a less-than-good outcome. All of this is not to disparage the value of social software, but to emphasize the need to balance functionality and security.[43] Serious, sophisticated applications of risk management (vice risk avoidance) are increasingly critical in any environment, and national security leaders and employees at all levels must learn to understand relevant trade-offs.

---

[40] See, for example, Priscilla Clapp and Morton Halperin, with Arnold Kanter, 2007, *Bureaucratic Politics and Foreign Policy*, Brookings Institution Press.

[41] The DEFCON convention, http://www.defcon.org/, is held every summer in Las Vegas, NV, and bills itself as "the largest underground hacker convention in the world," bringing together talented people with diverse viewpoints and often highlights serious security issues, and DEFCON typically includes more than 80 presentations in four or five parallel tracks.

[42] For example, SATAN (Security Administrator Tool for Analyzing Networks), a software program that claims to "identify weaknesses in just about any network connected to the Internet," was discussed as a way to exploit social software vulnerabilities.

[43] An interesting new perspective on this is an interview of MG Michael Oates, commander of Army Task Force Mountain, in an interview: "Blogging General Reaches Out to Troops, Blows Off Security Fears," Noah Shachtman in *Wired*, http://blog.wired.com/defense/2009/01/tf-mountains-so html. See also http://www.army mil/-news/2009/01/09/15633-commander-connects-online-with-soldiers-families-spread-around-the-world/.

On balance, care must be taken that data not be so guarded that the overall mission itself can not be executed. Commanders and policymakers must seek to achieve *mission assurance*—the ability to perform a mission irrespective of the level of attack suffered, rather than just *information assurance*. The downside of not sharing information must be articulated along with the risks of sharing. Very importantly, new oversight paradigms need to be introduced to keep traditional network security responsibilities and accountability mechanisms from choking off innovative approaches through criticisms from organizations that have not yet adapted to the changing paradigms. They also may be constrained in their ability to change by out-of-date laws and regulations that they are charged with enforcing. They could become allies in adopting new approaches, if engaged properly.

3.4 Infrastructure

The size and diversity of the USG, as well as frequent changes of supervisory personnel, make it hard to maintain a coherent ICT infrastructure. DOD alone has over 2.5 million non-embedded computers. In addition, after years of outsourcing, and the relatively senior age of USG managers, it is hard for government supervisors to keep pace with innovation in the larger marketplace. The Office of Management and Budget (OMB) e-gov initiatives have made a positive difference over several years, and the new Comprehensive National Cyber Security Initiative (CNCI) should continue the high level attention. But the integration and modernization of government ICT systems will remain a challenge.

3.5 Employee Demographics

Government employee demographics differ from many non-government organizations because of differing hiring processes, public service motivations and incentives, and effective tenure. In combination, these factors rarely encourage cultural change and the adoption of cutting-edge technologies. A related issue is how those who are motivated to deliver social software can actually provide products that will be understood and be used in productive ways by government end users, who are likely to be very different from the social software evangelist or developer.

Retaining quality employees, particularly those from the collaborative, creative class of largely late-Generation X and Generation Y people who use their passion and intelligence to mix work and play seamlessly, is a very important, constant struggle for the government.[44] In particular, this is because government service is slower to adapt to change than these "digital natives" are, and often offers less financial incentive than the

---

[44] Maxine Teller, a consultant to OSD Public Affairs, reports that 95 percent of digital natives use social software, 97 percent play video games, 75 percent use text messaging, 48 percent have web-based photos, and 28 percent have personal blogs.

private sector in many cases as well. Recruiting the right people into government to accomplish the missions laid out in this paper is only the beginning.[45]

3.6 Administration and Political Changes

Constant reshuffling and reorganizations at relatively senior levels of bureaucracies and inconsistent use of terms across agencies further complicate the long-term implementation of creative agendas.

3.7 Budget and Resource Restrictions

Despite the Defense Department's 6-year Future Years Defense Program (FYDP) and the size of the DOD budget, the vicissitudes of the appropriation and authorization processes make it very hard to sustain long-term budget cohesion within the USG, even for major acquisition programs that have strong congressional support. Also, because budget justifications, reports, requests, and decisions are made on a cyclic basis, the work put into this process takes valuable time away from other activities.

---

[45] A recent example of this occurred within NASA's CoLab program, http://colab.arc.nasa.gov/, a new initiative designed to serve as an internal NASA consultant, a bridge between NASA and the public to form collaborations. At the end of July 2008 they hired a well-known and highly regarded social media personality, Ariel Waldman, from the Silicon Valley company Pownce, http://arielwaldman.com/2008/07/28/exciting-news/. Three months later, she publicly resigned, http://arielwaldman.com/2008/11/03/update/. Briefly, her NASA contractor employer did not allow use of social media networks for her to do her job…on social networking. Waldman says, "The policies and mindsets are written such that it makes use of Twitter akin to playing Solitaire at work."

4. Social Software Usage on the Global Stage

Social software has implications for global security and stability. For example, emerging social software can rapidly and effectively rally people to causes and action. Internet technology has long been used by young people to mobilize to protest government policies, including 2004 protests in Ukraine and those in Belarus during 2006. In addition, communication among decentralized entities such as rebel, criminal, and terrorist networks has security implications.[46]

Social software has affected recent events around the world, including: the narco-terrorism of Colombia, the Russia-Georgia conflict, Facebook-organized political dissent in Egypt, the terrorist attacks in Mumbai, and the Israeli military action in Gaza. There is a clear progression of increasingly sophisticated use of social software in these situations. At the least, our military, intelligence, and diplomatic arms should be aware of the situations described below, understand the role of social software in global security and stability affairs, and be able to monitor or influence events as they unfold.

4.1 Counter-Rebellion Against the FARC Rebellion (2008)

In January 2008, Oscar Morales of Colombia started a Facebook group against the revolutionary guerrilla group FARC (*Fuerzas Armadas Revolucionarias de Colombia— Ejército del Pueblo*). What began as a group of young people venting their rage at the FARC on a website ballooned into an international event called "One Million Voices Against FARC" with the goal of destroying the FARC. Although the demonstration and associated rhetoric was controversial in many quarters, more than a million people turned out in more than 40 countries on February 4, 2008, one month to the day after the initial post, with relatively little involvement of the Colombian government. Here we have an early example of social web technology significantly mobilizing a large number of people with national security implications.

4.2 Russia-Georgia Conflict (2008)

During the Russia-Georgia ground conflict in 2008, a cyberwar of freelance hackers targeting state-run information websites was also being waged.[47] (Similar occurrences,

---

[46] "How Web 2.0 Has Changed Armed Conflict Forever," http://blogs.nyu.edu/blogs/age282/zia/2009/01/idf_use_of_web_20_represents_f.html; "How Terrorists May Abuse Micro-Blogging Channels Like Twitter," http://www.dhanjani.com/blog/2008/12/how-terrorists-may-abuse-microblogging-channels-like-twitter.html, "al-Qaida-like Mobile Discussions and Potential Creative Uses" (U.S. Army OSINT Draft FOUO), http://www.fas.org/irp/eprint/mobile.pdf.

[47] "Cyberwar 2.0 in Georgia," http://blogs.ft.com/techblog/2008/08/cyberwar-20-in-georgia/. See also: "Project Grey Goose, Phase I Report," (http://www.scribd.com/doc/6967393/Project-Grey-Goose-Phase-I-Report). Project Grey Goose is a non-USG Open Source Intelligence (OSINT) initiative launched on August 22, 2008 to examine how the Russian cyber war against Georgia was conducted.

sometimes called "cyber riots," occurred in 2001 across the Pacific between often unidentified activists in the United States and China in relation to the forced landing of the U.S. EP-3 reconnaissance aircraft on Hainan Island, and in 2007 during the cyber conflict between Russia and Estonia.)[48] Real-time "citizen journalists" provided excellent text and visual information about the Russia-Georgia conflict via sites like Twitter and Flickr, a popular, free photo-uploading and sharing site.[49]

4.3 Facebook Versus the Egyptian Government (2008)

In a show of "cyberactivism" in late March 2008, two Egyptian citizens launched a pro-democracy Facebook group to protest their government's policies, which include not allowing groups of five or more people to gather without a permit. Within about a week the group had attracted 40,000 members.[50] On April 6, 2008, the group had a protest of political dissent, and posted photos online of the violence that ensued. State security was taken by surprise by the number of participants. In effect, online social movements have changed the dynamics of political activism. Possibly as a result, Syria has recently blocked use of Facebook by its citizens.

4.4 Terrorist Attacks in Mumbai (2008)

The 2008 attacks in Mumbai unfolded online in real time, and the mainstream media (and in effect the world) got a first-person, eyewitness view.[51] Twitter streamed information and images during the terrorist event at such a rapid pace that mainstream media simply used footage without attribution and independent fact-checking. Hearsay and assumption also played a strong role in the information flow, and to some extent, "trust but verify" was suspended in favor of speed. While rapid, first-person intelligence via these new communications is valuable, there is the very real possibility of exploiting such streams to promote misinformation, particularly if decisionmakers do not understand the technology well. From incidents like this, the very limited time that governments have to respond effectively in crises where social software is part of the information flow is becoming readily apparent.

---

[48] See Kenneth Geers, "NATO Cooperative Cyber Defence, Centre of Excellence," Tallinn, Estonia (presentation), http://www.blackhat.com/presentations/bh-jp-08/bh-jp-08-Geers/BlackHat-Japan-08-Geers-Cyber-Warfare-slides.pdf

[49] "Flickr Group: Russians Out of Georgia," http://www.flickr.com/groups/russiansoutofgeorgia/.

[50] This is extremely large by Facebook standards—or by any standards. Facebook is the third most-visited site in Egypt, behind Google and Yahoo. About 1 million Egyptians use Facebook, or approximately 11 percent of the online population. See "Cairo Activists Use Facebook to Battle Regime," http://www.wired.com/techbiz/startups/magazine/16-11/ff_facebookegypt.

[51] "Citizen Journalists Provided Glimpses of Mumbai Attacks," http://www.nytimes.com/2008/11/30/world/asia/30twitter.html?_r=2.

4.5 Israeli Incursion in Gaza (2008-9)

During conflict in the Gaza Strip during December 2008 and January 2009, Israel set up an official Twitter stream based in its New York consulate.[52] In a very sophisticated manner, Israeli operators used the latest in Twitter tools, techniques, and lingo, including a two-hour "Twitter press conference" on December 30. Other ongoing Web 2.0 multimedia includes a blog and a YouTube video channel showing everything from air strikes to children receiving medical attention.[53] No doubt, this is an effort to shape world opinion about their recent actions. Hackers also were active on the Hamas side of the battle.[54]

4.6 Pakistani Chief Justice Protest (2009)

A state of emergency including blockage of private television channels was imposed on Pakistan in November 2007. Despite only 17 million Internet users in a population of over 150 million people, citizen journalists, lawyers, and activists began using social media to document their government's actions.[55] Technologies employed included blogs, chat forums, YouTube, and Twitter. In March 2009, this largely virtual social network of activists was enlisted in an ongoing effort to get the nation's Chief Justice reinstated, and contributed to its success.

4.7 Coup d'État in Madagascar (2009)

Civil unrest and political confrontation in Madagascar led to the resignation of its democratically-elected President, Marc Ravalomanana, in March 2009. Madagascar opposition leader Andry Rajoelina assumed leadership despite criticism from the country's courts and the international community. At one point, rumors circulated on the microsharing site Twitter that Ravalomanana was seeking refuge inside the U.S. Embassy in Antananarivo.

Concerned that the embassy might be attacked by opposition supporters, the U.S. State Department, via its preexisting "Dipnote" Twitter account, sent two "tweets" to dispel the rumor.[56] Text #1: "We are aware of media reports that President Ravalomanana of Madagascar is seeking sanctuary at the U.S. Embassy in Antananarivo." Text #2:

---

[52] "How the Israeli Consulate Brought the State to the People," http://rwy.blogspot.com/2008/12/how-israeli-consulate-brought-state-to.html; "Why Israel's Twitter Experiment Flopped," http://comops.org/journal/2009/01/12/why-israels-twitter-experiment-flopped/.

[53] The blog is called IDF Spokesperson, http://idfspokesperson.com/.

[54] Israel/Hamas Battle Goes Web 2.0, http://arstechnica.com/news.ars/post/20090105-israelhamas-battle-goes-web-2-0.html.

[55] "Following Pakistan's Protest March, Another Trail of Twitters," by Huma Yusef in *Christian Science Monitor*, http://features.csmonitor.com/globalnews/2009/03/16/following-pakistans-protest-march-a-long-trail-of-twitters/.

[56] See http://twitter.com/dipnote/statuses/1342746933 and http://twitter.com/dipnote/status/1342748135.

"President Ravalomanana has made no such request and is not in the US Embassy." According to the Associated Press, "Misinformation can have serious consequences, and when we saw the story breaking that way, we decided we had to do something about it quickly," said deputy State Department spokesman Gordon Duguid. [57] He added: "The situation was fluid and the embassy was open. We had to protect our people." This appears to be the first time the USG has used microsharing to avert a potentially serious crisis.

4.8 Election Protests in Moldova (2009)

After accusations of vote-rigging in Moldova's parliamentary elections, more than 10,000 Moldovans, mainly people associated with youth groups named Hyde Park and ThinkMoldova, protested against Moldova's Communist leadership on Tuesday, April 7, 2009.[58] This involved taking over government buildings (including the President's offices) and violently interacting with police who attempted crowd control. For some time, Internet service from Chişinău, the capital of Moldova, was cut off in response to the tremendous rush of firsthand accounts of the protests via Twitter and other mobile-enabled Web 2.0 services, according to media accounts. One of the leaders of ThinkMoldova described the initial effort as "six people, 10 minutes for brainstorming and decisionmaking, several hours of disseminating information through networks, Facebook, blogs, SMSs and e-mails." According to the latest news, social software continues to be used, some in English and much in Russian, to continue the conversation and possibly plan further action.[59]

4.9 China and Global Public Opinion

The Chinese Communist Party employs thousands of people to actively influence online public opinion.[60] These professional propagandists operate covertly in chat rooms, message boards, blogs, and so forth in order to advance party agenda items that may have U.S. national security implications. For example, they could impact efforts to limit the spread of avian influenza. Whereas such a campaign can be effectively shut down by a free press and subsequent negative public opinion, these checks and balances often are not effective in China since the government is not accountable to either entity.

---

[57] See "U.S. 'tweets' down embassy rumor," WTOP News, http://www.wtopnews.com/? nid=116&sid=1626750.

[58] "Protests in Moldova Explode, With Help of Twitter," by Ellen Barry for the *New York Times,* http://www.nytimes.com/2009/04/08/world/europe/08moldova.html.

[59] We note that this event is in-progress at the time of writing. The following article coves some of the increasingly sophisticated Web 2.0 technology being used in more detail. "Inside Moldova's Twitter Revolution," *Wired Danger Room,* http://blog.wired.com/defense/2009/04/inside-moldovas.html.

[60] "How China's '50 Cent Army' Could Wreck Web 2.0," http://itmanagement.earthweb.com/columns/ article.php/3795091/How+Chinas+50+Cent+Army+Could+Wreck+Web+2.0.htm.

5. Social Software and Security - Recommendations

In one of its first significant acts in January 2009, the Obama administration issued a directive calling for more transparent, participatory, and collaborative government.[61] DOD and other agencies with national security roles could benefit from the collaborative, distributed approaches promoted by responsible use of social software.

Our recommendations are not exhaustive, but this paper outlines a general foundation and strategy for moving forward with the incorporation of social software into national security missions. In addition, many of these recommendations have application to USG and other entities not dealing with national security issues.

5.1 Lead by Strongly Supporting Social Software

The success of anything novel proposed in this paper depends on strong, enduring leadership that encourages change. While the Obama Administration and certain isolated leaders have shown very positive signs of this, change must happen across entire agencies.[62] People must have the freedom to experiment, the incentive to innovate, and a leadership they can trust and emulate. Senior leadership should promote information sharing for adaptiveness and creativity. Current policies that prevent adaptation and innovation need a fresh look.

5.2 Analyze the Balance Between Security and Sharing

Information security concerns are non-trivial, particularly for DOD missions. This has led to a general sense that security officers will say "no" to experimenting with social tools on DOD computers. While some measure of "no" is reasonable, there is a point at which a mission can be hurt by strictly enforcing such draconian approaches that it keeps government from taking advantage of social tools that adversaries and other counterparties are using. This situation can also lead to DOD employees using insecure social software "workarounds" to accomplish their missions.

This does not only apply to the battlefield. In one anecdotal example, a public affairs group is charged with promoting military videos online using YouTube, but simultaneously cannot access YouTube from their work computers. A balance between too much security and too much sharing must be struck. Social software needs to be

---

[61] White House Memorandum for the heads of executive departments and agencies, "Transparency and Open Government," January 21, 2009, http://www.whitehouse.gov/the_press_office/TransparencyandOpenGovernment/.

[62] For example, U.S. Coast Guard Commandant Admiral Thad Allen introduced his social media initiative for the entire Coast Guard, using YouTube video, http://www.youtube.com/watch?v=vdEAY1XLapQ.

looked at within the framework of risk management and mission assurance. Sometimes, the risk is worth the reward.

5.3 Create a Culture of Social Software Experimentation

The use of Twitter by the U.S. State Department during the coup d'état in Madagascar occurred only because that agency was experimenting with the social tools before they "needed" them. There had been no requirement for or procurement of social software, but Twitter (in this particular case) proved extremely useful. Experimenting with emerging social software, using blogs and other online forums to read about topics of interest to an office's mission, and having publicly available contact information via the LinkedIn professional social network, GovLoop, or other sources are all steps that can largely be taken immediately by many USG employees and embedded contractors.

It is not easy to assign one office or job function to answer the question of, "who should have the lead with social software usage?" (figure 3). This is true in both the public and private sectors. A recent article[63] posits that social software's hallmarks are community, dialogue, and partnership, and it is simultaneously part of three worlds that we live in: the physical world (television, radio, print, and human interaction), the digital world (non-interactive websites, email, search engines and portals, etc.), and the virtual world (online interactive spaces like social networks, blogs, podcasts, and microsharing platforms with avatars and potential anonymity).

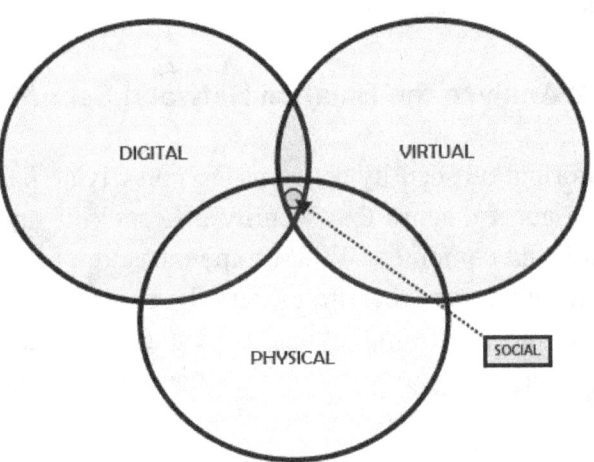

USG employees who will excel at social software, and specifically social media, are not obviously experts in public speaking, information security, public affairs, or web design—they tend to be generalists with a unique combination of knowledge about the physical, digital, and virtual information worlds, integrating them into a practice of social influence. Similarly, solving the problem of who utilizes social media for an

Figure 3: Knowledge of social software and its uses comes from an intersection between the physical, digital, and virtual worlds.

[63] "Who owns social media?," Joseph Jaffe, AdWeek, http://www.adweek.com/aw/content_display/community/columns/other-columns/e3i781c3e0a48f6c1c28c8684899749ce3d.

office or agency is likely to be more difficult than just hiring a public relations firm, a digital design firm, or a large IT contractor.

5.4 Network Government With the New Digerati

Whereas organizations like DARPA to a large extent led the first information revolution through public sector research funding, Web 2.0 technologies are mainly developed by small startup companies outside Washington, DC that receive no public funding and largely do not do business with the public sector at all. When the authors would travel and meet with people from Web 2.0 companies outside Washington, DC during 2008, most were surprised that someone from the USG was at their event and interested in their business. DOD and other USG components should develop better firsthand connections with these startup companies, venture capitalists, and bloggers in that sector. Enduring connections will primarily result from attending "their" conferences and other networking events in person, particularly in areas like San Francisco, Los Angeles, New York, and so forth.

5.5 Prepare to Discard Some Legacy Systems and Processes

Learning and mastering social software of any flavor is a time investment—practice makes perfect. So, if social software is incorporated into USG work as an add-on and not a replacement, training and therefore usage will be difficult. To some extent, legacy ICT systems, office processes and other normal ways of conducting work must be replaced by new ways of doing business, including social software tools. Just as, to some extent, television replaced radio (but did not eliminate it as a medium) and mobile phones replaced landlines (but again, did not eliminate them), some kinds of social software (like networking and microsharing platforms) have the potential to partly replace email and other traditional modes of communication during daily work.

As an example of the drawbacks involved with non-replacement of legacy approaches, the wiki-based encyclopedia called Intellipedia within INTELINK is often used as a shared platform for exchanging and integrating raw intelligence horizontally among the different IC agencies. However, information is often subsequently ported from the shared platform to a proprietary, agency-specific platform and used in a vertical manner within a "cylinder of excellence." Replacing stovepipe platforms with shared ones can be easier, faster, and more effective, and allows everyone to view and develop the same "Living Intelligence" simultaneously, while it grows and the analysis evolves.[64] Despite this,

---

[64] 'Living Intelligence' (also known as 'Purple Intelligence') is a term popularized by Chris Rasmussen, a social software enthusiast from the National Geospatial-Intelligence Agency. For more info see http://fcw.com/articles/2009/03/11/fose-web-20.aspx. In the military, "purple" means multi-service or joint, and "living intelligence" is a new metaphor for when an intelligence topic is constantly updated. This stands in contrast to the more typical snapshot notion of finished intelligence. The IC still produces about 50,000 reports a year, often lengthy and redundant. See http://www.denverpost.com/search/ci_4216851.

while some units have re-aligned reward structures and moved their "customers" into newer and more transparent and conversational processes and leaders in the IC have pronounced the importance of moving from a mentality of "need to know" to "need to share," the majority of the IC still upholds the idea that community-based horizontal collaboration is "suspect" compared to individual agency vertical vetting.

5.6 Empower Some Individuals to be Authentic

Social software can be very empowering for individuals—it can promote messages outward, filter news stories via word-of-mouth voting, drive traffic to agency blogs, increase people's public profiles, and many other things. But individuals often need permission to use it in this manner. The essence of social software is personal communication among persons acting authentically (in the sense of acting more individually and less organizationally) and transparently,[65] yet within the governmental bureaucracy there can be negative consequences to behaving in this manner. Groupthink should be discouraged as much as possible; there should be no "fear of attribution."

Increasingly, the private sector has seen a change in their human resources, whereby employees' personal and work lives are intertwined and government cannot avoid this trend. They have increasingly been coming to terms with people using, say, Twitter for work-related activities during the day, all the while incorporating personal quips, family photos and so forth. Similar situations occur with work phones and email accounts. Generally, problems are behavioral and not technological—social software is an ICT tool that should not be held to a higher standard than similar but more traditional tools like email.

5.7 Unlock the Government's Cognitive Surplus

The USG writ large has millions of employees and contractors, and there are countless experts or knowledgeable amateurs on government-related issues who work in many locations on many things. Often, many in the USG think that all the experts on a topic are within the offices working on that topic. But this is anything but the truth—wisdom is scattered in tiny pieces within agencies, and across the USG. Similarly some knowledge is inside the USG within Washington, DC and some is contained in the minds of employees and contractors around the country, and indeed the world. Social software platforms like microsharing, and constructs like metadata tagging help to connect this knowledge in relatively simple ways, empowering people to contribute and allowing interested parties to search and discover information helpful to them.

Purple intelligence can help reduce the amount of duplication by moving the review process into the same space where the collaboration takes place.

[65] Social software marketing expert Gary Vaynerchuk calls this being a "RAT"—real, authentic, and transparent.

Simple social software platforms for enterprise microsharing like Yammer and Present.ly, promote social networking and serendipitous information exchange. The IC, NDU, and other USG agencies are experimenting with Yammer and similar tools, but their value scales with the number of users, which currently is small. Other potentially useful tools for unlocking intra-organizational cognitive surpluses include internal prediction markets, which have been used by DARPA and Google, among others.[66]

5.8 Envision Citizens as Communities of Conversations

An increasing number of people are using social software in their personal lives, whether they realize it or not. They look up information in wikis, comment on blogs, chat with friends, and so forth. This incredibly open and inexpensive information sharing has resulted in people forming social networks around topics of interest, and having conversations about them. So, citizens are no longer willing and empty vessels waiting to unidirectionally receive news from a USG press release, the 6:30 pm network news, or the morning edition of *USA Today*.

Citizens are continuously engaging in conversations about topics of interest to DoD and indeed every part of the USG—military recruiting, foreign policy, infectious disease, education, the environment, and many, many others. Within the law, the USG can use social software to listen to those conversations, engage in conversations, and perhaps then try to influence people by providing helpful information while being transparent about where they work and what their agency's mission(s) and policy(s) are. Otherwise, since the conversation is happening anyway, the government will be left out of it. The notion of citizens constantly having conversations of interest to the USG applies not only to U.S. citizens, but also to people in many countries around the world.

5.9 Create Return on Engagement Through Indirect Influence

Through social networks people often develop online relationships before meeting in person, and keep relationships intact between sometimes distant meetings in person. These online relationships can be invaluable. In the case of Colleen Graffy from the State Department (see Section 2.4.1), her online relationships developed via microsharing professional and personal information on Twitter developed a sense of familiarity with foreign journalists before she met them in person—comfort was more easily established, time was saved, and more than likely a significant impact was left. Moreover, when official interviews end, a relationship can continue online at low effort.

---

[66] Prediction markets are a variation on financial markets that are also called information markets and event futures. See "Prediction Markets" by Justin Wolfers and Eric Zitzewitz, Journal of Economic Perspectives 18: 107–126, http://bpp.wharton.upenn.edu/jwolfers/Papers/Predictionmarkets.pdf.

Commonly, people ask what the return on investment (ROI) is from using social software—for example, what is the value of five tweets per day? How did that make more profit, or help my government mission? The answer is that in many cases, the "investment" has not changed—official phone calls, travel, meetings, press conferences, and so forth are the true investment of time and money. But perhaps a novel term is appropriate here - "return on engagement" (ROE). In principle, ROE can be very high. Pre-engagement of specific people or general groups of people having conversations about your mission can pay dividends when it is time to make an investment using more traditional means. But by pre-engagement, a DOD employee can indirectly influence people by talking about who they are, what they read, and what their professional goals are, yielding ROE when done correctly (see figure 4). ROE can enhance ROI:

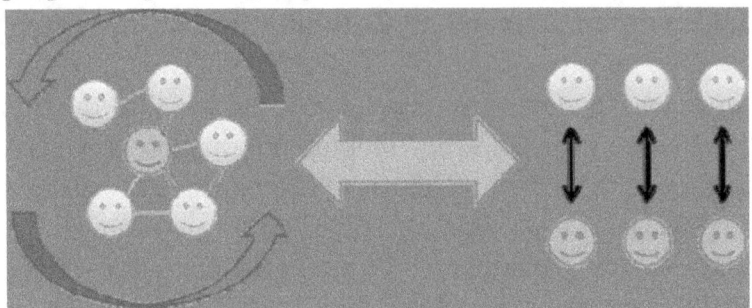

Figure 4: Return on Engagement (ROE) interacts bi-directionally with Return on Investment (ROI). In this model, ROE via social networking can enhance, but not replace, ROI from more traditional activities.

whereas in the previous industrial/broadcasting era information was power, in the new collaboration era, information *sharing* is power.

5.10 Develop Modern "Brands" and Market Them

DOD components, and even offices within them in some cases, should develop engaging, trusted public voices independent of the public affairs office, which is often seen as being a bland microphone for "the official" positions of the government. Through extensive personal engagement with citizens, attendance at public events, appearances on "lite" talk shows, and other avenues like blogging, microsharing, or engaging in virtual worlds, personable representatives of DOD can work on behalf of the government among the people, but also work on behalf of the people within the government. Through such engagement involving information sharing, the most engaged and sharing person in a community often becomes the most trusted.

The concept of marketing government brands is not as strange as it may initially sound. After all, the volunteer-based military services have recruiters whose job is specifically to engage with people and "sell them on a brand"—not to mention television commercials and other forms of traditional advertising on this topic. Partly using social software, that concept can inexpensively be extended to other offices and missions. Persons who might act as DOD brand representatives should:

- be as authentic and transparent as possible;
- be leaders regardless of their place in the government hierarchy;
- understand interacting with people in physical, digital, and virtual spaces;
- be publicly accessible via multiple media channels;
- be catalysts that make change happen through their intelligence and presence;
- conduct community-based research whose goal is to understand citizens better;
- direct open discussions about their DOD office's mission and brand;
- add value by sharing information with their community of interest;
- embrace both positive and negative feedback from the community;
- understand public sentiment and acknowledge valid criticisms;
- make suggestions so DOD can provide better services to citizens; and
- be authentic and admit that they make mistakes from time to time.

5.11 Answer Questions Within Specific Areas of Responsibility

Here we briefly describe the key questions about social software that should be answered by various offices in DOD depending on their missions and responsibilities. This section is based on the organization of the Office of the Secretary of Defense (OSD) but also could apply to the Joint Chiefs of Staff, the combatant commands, individual military services, and other agencies and groups working on national security issues. The list of offices and questions is not exhaustive, but this section highlights some key issues that need to be examined by policy researchers, in workshops, or during more formal decisionmaking processes within DOD.[67]

Area 5.11.1: Policy an d General Counsel. The key question about social software for decisionmakers in policy and government law and regulations is: What are the DOD policies for incorporating social software into daily work? New policies, doctrine, and other rules should encompass the four general functions of social software for the government described above: Inward, Outward, Inbound, and Outbound Sharing.

Area 5.11.2: Personnel, Readiness, Training, and Education. The key question about social software for decisionmakers in personnel, readiness, training, and education is: Who can use what social software for which purposes, and how should they be trained and prepared to use it? Other important questions include: What are the individual incentives to learn to use social software, and use it well? What individual responsibilities are there for cyber security when using social software on government computers in a Web 2.0 environment?

---

[67] This section might be seen as offering guidance to a more formal DOTMLPF (Doctrine, Organization, Training, Materiel, Leadership, Personnel, Facilities) analysis of social software and national security.

Changing military forces and their civilian support within DOD with regard to social software will require new or different training and education. As appropriate, mid- and senior-level officer schools should teach about and use social software. Such training should be related to functional specialty (information operations vice public affairs, and so forth). Like some aspects of cybersecurity, DOD could require online social software courses of all on-site personnel using USG work stations, to promote education about benefits and costs to using social software in the government environment. Finally, even young, newly enlisted recruits should learn about policies for being members of private social networks, using microsharing, and other popular social software inasmuch as it could compromise operational security or pose other risks to DOD.

Area 5.11.3: Acquisition, Technology, and Logistics. The key question about social software for decisionmakers in the areas of acquisition, technology, and logistics, to include DARPA, the Defense Technical Information Center, and others, is: What social software do we need for what DOD missions, where do we obtain it, and how do we adapt it for government use? Another question is, what research related to social software is required to solve national security problems that current software cannot?

Area 5.11.4: Intelligence. Two key questions about social software for decisionmakers in this area, to include the National Security Agency, the National Geospatial-Intelligence Agency, and the Defense Intelligence Agency, are: How can DOD best incorporate social software into the performance of intelligence and counterintelligence missions? How can DOD best understand and articulate the intelligence issues concerning the global use of social media and the counterintelligence issues related to operational security and force protection? How can IC social software tools be integrated with non-intelligence networks? There are many possible applications of social software to the field of intelligence, and the IC has made significant inroads, but there are many unresolved issues.

Area 5.11.5: Networks, Information Integration, and Chief Information Officer. The key question about social software for decisionmakers working on these missions is: How does DOD manage risks to information security while assuring mission success via incorporation of social software? A new balance between restrictions and opportunities should be reached. Unless some blessing from security experts is received, the inherent conservativism of government will make it too easy for people to say "no." Care should be taken that limits not be so restrictive as to preclude the benefits of these important tools.

To reduce visceral, security-related objections, we propose that social software (at least some of the most useful or popular sites—see our table) should be binned into four broad categories by DOD and IC security professionals: (1) suitable for general use with appropriate guidelines, (2) acceptable for use outside firewalls, (3) acceptable for use within protected enclaves, and (4) not suitable for government use. Results from experiments testing the security vulnerabilities of various social software platforms will influence local rules and personal behavior guidelines at different DOD units with different responsibilities.

Area 5.11.6: Public Affairs. The key question about social software for decisionmakers in the area of public affairs is: How can DOD best utilize social software to communicate and engage with the public? Does audience interest in DOD-related topics match mainstream news coverage of those topics and, if not, how can emerging media forms be used to bridge those gaps?

DOD should adopt new mindsets for public outreach, and look widely in government and industry for social technology best practices.[68] DOD should benefit from enhanced public interactions that involve multi-directional, multi-media engagement with citizens. They should also benefit by having engaging, public faces of DOD use social software to market DOD brands and therefore further the national security mission.

To some extent, DOD is already doing so, with innovative programs like Bloggers' Roundtable and Pentagon Talk Radio's "Armed With Science" podcast.[69] An increasingly fragmented media, combined with a technologically empowered public within which everyone has the ability to act as a collector, analyst, reporter, and publisher leaves more opportunities—and pitfalls—to engage than ever before. These trends in media and technology appear to be accelerating, both among mainstream press and among bloggers and similar "amateur" journalists.

---

[68] Government 2.0: How Social Media Could Transform Gov PR, http://www.pbs.org/mediashift /2009/01/government-20-how-social-media-could-transform-gov-pr005.html.
[69] See http://www.defenselink.mil/blogger/index.aspx for the near-daily Bloggers' Roundtable, and http://www.blogtalkradio.com/stations/PentagonRadioNetwork/ArmedwithScience for the weekly Armed With Science online radio show.

6. Conclusions

Since April 2008 the Center for Technology and National Security Policy at the National Defense University (NDU-CTNSP) has been engaged in research on social software in support of DOD policymakers. We call this research project Social Software and Security (S3). This preliminary "net assessment" on S3 draws on extensive connections among influencers and thought leaders throughout USG, non-profit entities, and the private sector. Goals of the project include:

1. conducting an inventory of available social technologies
2. tracking global government social technology case studies
3. identifying impediments to social software use in DOD
4. engaging with private sector experts for informal advice
5. providing advice to DOD senior leaders on aspects of S3

These objectives are in various stages of progress. We have developed a fairly complete inventory of commercial off-the-shelf (COTS) social software and their features, much of which has been made available via a self-organizing public community website, http://sniki.org/ ("Sniki" is an amalgamation of "social networking" and "wiki"). The authors have contacted information assurance (IA) professionals from DOD, in particular NSA, to ask for help in binning the social software products as described above. By extensive networking throughout the diverse communities of traditional ICT, social software startup, media, marketing, and defense we have established a knowledge base and network of individuals involved with different parts of the overall S3 slate of issues. Through conversations and online and print writing, we have begun advising senior decisionmakers and their staffs on how to implement social software in their organizations.

Over half of the current members of the U.S. armed forces were born after 1980, and many civilian DOD employees and contractors belong to that same generation. They are digital natives who have seemingly always had mobile phones and do not remember card catalogs or a world without the Web. The people of Generation Y are more likely to get news from the Internet than a newspaper, and perhaps more likely still to have that news filtered through the word-of-mouth of their colleagues, friends, and trusted acquaintances.

At the same time, DOD middle management and senior leaders come from a different, but equally valid vantage point. To some extent, there is a cultural disconnect regarding how people interact with each other and share information, not only within DOD but also within many other organizations, public and private.[70] Many people of the Baby Boom

---

[70] Misti Burmeister, *From Boomers to Bloggers*, Synergy Press, 2008.

generation and even early Generation X range from somewhat unknowledgeable to highly skeptical about the professional value of social software, while others have seen it used and understand its importance, but find learning about such a broad new topic overwhelming.

On balance, DOD needs to appreciate that social software exists, is becoming increasingly popular, and has empowered people to self-organize outside government and other major institutions without permission, endorsement, or encouragement. DOD needs to be prepared to not only research, build, and/or acquire social software tools, but also to be prepared to educate its workforce about how to use them, and why.

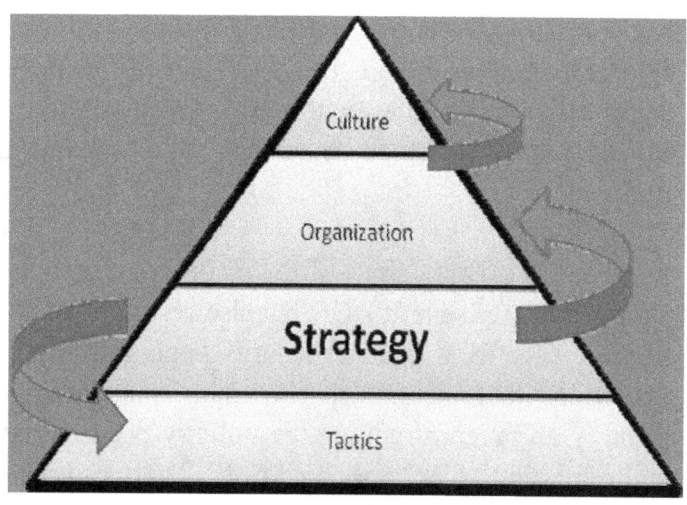

Figure 5: A social software strategy for DOD (government) ideally precedes formal action. This strategic policy then influences both specific tactics used to meet objectives, and also changes in organization and culture.

But, most of all, DOD needs a strategy (figure 5) that envisions how social software can be used to complete its missions better, and even envision new missions and goals that were not before possible. Such a strategy will guide not only the specific social software tactics that are deployed to tackle problems, but also—and this is more of a challenge—organizational and cultural changes necessary to transform DOD into a body where information flows more freely.

Experimentation with social tools is educational and should be encouraged; however, experimentation alone is tantamount to tactics in search of a strategy, or as the "new marketing" firm Crayon tells its clients, "You have solutions in search of a problem." As in the private sector, starting with a strategy for using social software that includes vision and planning will form a foundation for both downstream tactics and upstream organizational changes and cultural buy-in.

Based on the examples given earlier in this paper, social software is clearly not a bunch of "nonsense for kids." To the contrary, it is an important information sharing enabler between individuals within government, between government employees and communities of interest, between researchers and government data, between the government and its citizens, and between governments of different countries.

Nevertheless, important issues must be addressed before pressing forward with DOD adoption. First, strategic mistakes could have unintended consequences across the diverse axes of national security writ large: military, information, diplomatic, legal, intelligence, financial, and economic. Second, tactical mistakes can, among other things, open significant security holes within ICT systems. And third, organizational mistakes can lead to social software usage being more of a wasteful time-sink than a benefit to ongoing missions.

Developing a DOD strategy for incorporating social software into its diverse missions should not only benefit DOD but also result in positive downstream effects on other parts of the greater U.S. national security apparatus. With an increasing emphasis on military operations that include peacekeeping, humanitarian efforts, domestic situations, and stability and reconstruction, the military needs to be adaptive. But it cannot handle the entire mission alone, nor should it.[71] There are many applications of social software to national security, including missions that fall primarily under the State Department, the U.S. Agency for International Development, the Department of Homeland Security, and so forth, where DOD serves as a supporting, but still important, partner. An overall strategy should involve using social software as an interagency connective tool and a human intellect force multiplier.

One current area within the greater scope of national security missions in which social software may have a large, immediate effect is in aspects of communication of American goals and values in other countries. Colleen Graffy's use of Twitter as an ROE tool for public diplomacy is but one example. Social software is useful for general networking within communities, listening to real-time conversations on topics of interest, identifying emerging influencers within micro-niches, providing mechanisms for combating negative viewpoints, and measuring public sentiment that may influence internal policy and program guidance. This dovetails with the notion of reviving the independent U.S. Information Agency, whose purpose was to support U.S. foreign policy and national interests abroad by conducting international educational and cultural exchanges and communicating U.S. messages, in a more powerful capacity.

In a broader sense, social software may help the overall USG mission—in what is commonly termed "e-Government" or "Government 2.0"—of national governance, to include all three branches of government. With an ever-increasing size and complicated

---

[71] Hans Binnendijk and Patrick M. Cronin, eds., *Civilian Surge: Key to Complex Operations*, National Defense University Preliminary Report, 2009; Hans Binnendijk, "At War But Not War Ready," *Washington Post,* November 3, 2007, http://www.washingtonpost.com/wp-dyn/content/article/2007/11/02/AR2007110201725.html.

nature of government and the issues with which it deals, the very notion of checks-and-balances is changing. Information flows at a higher velocity than ever, yet (partly because of a real-time news cycle[72]) time scales for informed decisionmaking are shrinking—what used to be decided in weeks or days now needs to be decided in hours. Moving forward in this new information environment, knowledge gained from a thorough understanding of social software and its applications to governance can enhance the current system of checks and balances. A system in which due deliberation in the absence of paralysis or discussion hijacking occurs within a limited window yet still allows countervailing opinions to be heard is ideal. Social software can help a rapid flow of information to be incorporated and digested easier using non-hierarchical means within existing hierarchical government decisionmaking structures.

Ironically, one of the most horizontal, self-organizing, and adaptable organizations on earth may come from within DOD itself—the human behavior that manages flight operations on aircraft carriers.[73] While there is a hierarchy of roles and a strict chain-of-command (starting in principle with the Captain and the "Air Boss"), within this hierarchy there is adaptation—no two carriers' flight deck personnel operate precisely the same way. Different ships have slightly different layouts and other peculiarities, and the sailors working in those conditions adapt appropriately, resulting in relatively few mistakes given the complex and dangerous nature of their tasks.

Warfighters in combat situations are very adaptable to changing environments. If these attitudes pervade decisionmaking on issues of policy related to social software and security, perhaps the answers will come from within.

---

[72] See lectures by BBC anchor Nik Gowing, for example,
http://ccw.politics.ox.ac.uk/events/archives/mt04_gowing.pdf.
[73] Gene I. Rochlin et al, 1987, "The Self-Designing High-Reliability Organization: Aircraft Carrier Flight Operations at Sea," Naval War College Review, http://www.fas.org/man/DOD-101/sys/ship/docs/art7su98.htm.

www.ingramcontent.com/pod-product-compliance
Lightning Source LLC
Chambersburg PA
CBHW081403170526
45166CB00010B/3194